God...The Infinite Lover

© *Edward J Hahnenberg, 2015*

Forward

Astrophotography combines two very challenging and expensive hobbies into one large hobby. It is without a doubt the most technically challenging and demanding kind of photography out there. We are shooting in extremely low light conditions very faint and distant objects with very long focal links and to top it off these objects are moving across the sky. Normal camera focusing doesn't work in low light nor is it sufficient to "just put it in on infinity" as even a 5° change in temperature requires the focus to be tweaked because of the contracting metal of the telescope. Critical focusing proves to be exceptionally challenging as it turns out that nothing is harder to focus on than a star. Ordinary daylight camera optics show every focal flaw, astigmatism, and every other conceivable optical aberration as there is no more challenging torture test to optics than a field of stars. Optics must be exceptional and thus very expensive. The telescope must track the object perfectly for hours on a rock steady base while the camera slowly and precisely collects ancient photons. A 1000 mm focal length and a 2° field of view is considered average or normal in astrophotography . At this kind of magnification there is little room for error of any kind. But with hours of exposure time in the dark in the cold with tired eyes and foggy brain there are lots of opportunities for error. Invest several thousands of dollars and you can begin to take some pretty nice pictures. Of course that is only after hundreds of hours of frustration as you learn the ins and outs of your equipment. Better and more expensive equipment makes things a little easier but you still need to be out in the dark under the stars to make it all happen and fix the inevitable things that won't quite work as planned.

Those of us that are into this hobby really need to have our heads examined. We spend many thousands of dollars to stay up all night miles from the nearest man-made light ,taking hours of exposures of objects often too faint to see through the telescope. And it isn't like we can sell our prize best pictures to recoup the huge costs involved better. Better pictures are easily and freely *available from the Hubble space telescope. Even those exceptionally rare individuals that have nice enough images to sell, considering how expensive the equipment is and how much time it takes to create a salable image it would be better off working as a greeter at Walmart than doing astrophotography. Rarely do we photograph anything new. The Hubble space telescope and other professional telescopes around the world have imaged everything interesting in the night sky hundreds if not thousands of times before at resolutions and quality that are simply technically impossible to match for amateurs.*

What could possibly be the appeal of this crazy hobby? I think that in part it is because it is so challenging that it is so appealing. It isn't for the impatient or budget minded amateur astronomer. This hobby takes a lot out of you, physically and mentally, emotionally, and especially financially. But so does climbing mountains and there are thousands of people who go out of their way to do that. Everest has been climbed hundreds of times, but people keep wanting to climb it. People spend tens of thousands of dollars and years of their life trained to stand on the summit themselves. Why? They want the challenge. They want to conquer something, and it isn't about the money. There is something about that for an astrophotographer as well. Sure the Orion nebula M 42 has been imaged thousands of times before, but I took this one!

I also think that there's something very sad about amateur astronomy in general and astrophotography in particular.

Time flies by for me when I get into my groove out in the dark and dawn comes a little too quickly. Under a canopy of stars, where the Milky Way shines so brightly over your head, you really get a sense of being on the earth spinning in space. It is often very quiet in nature, and things tend to move very slowly at night. It is a time to be still, look up and just notice. This experience isn't on TV or the Internet. This is the real thing, the universe, and if you really pay attention, you can feel your place in it. You are tiny beyond imagination, and yet you are also here and aware of it all. Like when you visit any great cathedral, it is always nice to bring home a little reminder of your experience there.

Astrophotography does it for me. 99% of my life is spent doing my daily stuff, handling responsibilities, making money, and raising my kids. This hobby gives me the opportunity to remember a greater context for my life and my place in the vastness of time and space. Astrophotography lets me steal a wisp of the vast mysteries of the night sky and bring them into the light of day to chair. Who wouldn't love that?

(Mr. Vedeler is an acupuncturist and amateur astrophotographer whose website is http://www.iso media.com/homes/cvedeler/ space.htm. Article printed with permission.)

Chapter I

Who is God?

It all began when I began to think about the infinity of space. I had been thinking about Aristotle, who, when he thought about the causality of things, realized that the world that he knew had to have a beginning sometime and somewhere. He reasoned that there was motion in the world. The further he thought, he realized that if he went far back enough time, he would come to a time when there was still motion in the world. The farthest back in time, he reasoned. there was no real beginning. So being the logical philosopher that he was, he began to take on the idea that there must be motion that had no beginning. . He calls this motion eternal.

In book 11 of his metaphysics, Aristotle goes into great detail explaining this. In this book, which consists of 12 chapters, Aristotle delves into many examples about the question of "being". He draws upon the principle of non-contradiction, saying that a thing cannot be and not be at the same time. As I was thinking about Aristotle's argument, I thought about the axioms in geometry. When I took my geometry course in the 10th grade, there were axioms that were accepted without proof. For example,a line was defined as having no beginning and no ending. A segment of a line has a beginning and an ending. If we think of as a line in our physical bodies, it, has a beginning and an ending.

I thought about the truth of mathematical statements, such as one plus one equals two. Was there ever a time when this was not true? Some may say that the truth was defined by human beings, and came into being only when humans recognized it as true. However, I don't agree with that. No matter whether humans determined what the definition of mathematical statements are, and no matter whether humans even existed on the planet, the truth that one plus one equals two was always true. This is an ontological truth. $1 + 1 = 2$ was true and always will be true, and thus has an eternal existence.

Now those who believe in a God accept the idea that God is eternal. Nothing in nature that we know about came into being on its own. All physical things have a beginning, and even the most durable physical things, it seems to me, will eventually lose their shape, size, weight, and so forth.

There are those that believe that at death, humans, in their material bodies simply cease to exist, except for the body decaying into molecular substances. Buddhism teaches this.

However in most human beings there is a desire for something beyond the grave. This manifests itself in religions and philosophies of various kinds.

In my upbringing, I had been brought up in a Judeo-Christian background. As a Christian, I believe in a Trinity of persons: Father, Son, and Holy Spirit. For many years I had believed that the

God of the Jewish people had chosen that race to be his favorite people. From that people, according to the prophets, Isaiah particularly, I thought it had been prophesied that a Savior would come from the root of Jesse. This is where the story of King David comes in. David was an extraordinary person in Jewish history. However, we know very little about him. He is sometimes described as a poet, a musician, and a military leader. In one of the books of the Old Testament David says: " Why should I will have all of this abundance of gold and other riches that God has allowed me to take from my enemies?" David wondered why he was able to live in luxury while the God who gave him these gifts, was present in a tent. So he thought it was his role to build God a temple. That, however, was his not his task. It would be another person… his son Solomon who would build the great temple.

It was Isaiah who prophesied that from David would come a Messiah. Now the Jewish people were looking for a Messiah at the time of Christ. Their concept of a Messiah was one who would free them from Roman control. But that was not

Christ's intent in coming to this earth. It was not to be a world leader. His task was to be a leader and Redeemer and remove the sin of Adam from the human race.

I have been an astrophotographer for many years. I have photographed many nebulae, star clusters, and galaxies. Our Milky Way galaxy contains upwards of 200 million stars. Some estimates are that there are even 600 million stars in our galaxy. Of these stars, there are many solar systems. None so far have planets that are habitable, although there is speculation that there are possibly hundreds of thousands of stars that have planets that may contain life.

One of the most spectacular images that I have photographed is known as "The eye of God", or the Helix nebula.

The Helix nebula which is as the constellation Aquarius lies about 700 ly away. Planetary nebulae, such as the Helix nebula are not only found in the Milky Way galaxy, but in many faraway galaxies.

In my experience as an astrophotographer, I have imaged many galaxies…even so far as imaging clusters of galaxies.

I could go on and on about my various astrophotography images, but that's not the purpose of this book. The book is about the infinite love God for his creation. But in order to understand his creation, we have to look at the physical universe around us. One of the wonderful tools that we have is astrophotography.

Astrophotography has given me a perspective of the immensity of the creation that we know now. Scientists estimate that this universe that we know came into being about 13.7 billion years ago. Imaging not only galaxies, but clusters of galaxies, has given me the perspective of living a good life on earth but kind of caught up in astronomical wonder.

Reflections:

There are many works in the Bible that speak of the universe God created. By some estimates, there are well over 40.

The book of Job, in chapters 38 and 39, contain the Lord's speech to Job and his accusers.

Chapter 38:

> *"Then the Lord addresed Job out of the storm and said: Who is this that obscures divine plans with words of ignorance? Gird up your loins now, like a man; I will question you, and you will tell me the answers! Where were you when I founded the earth? Tell me, if you have understanding. Who determined its size; do you know? Who stretched out the measuring line for it? Into what were its pedestal sunk, and who laid the cornerstone, while the morning stars sang in chorus and all the sons of God shouted for joy? And who shut within doors the sea when it burst forth from the womb; when I made the clouds its garment and thick darkness its swaddling bands? When I set limits for it and fastened the bar of its door, and said: Thus far shall you come but no farther, and here shall your proud waves be stilled!*

Have you ever in your lifetime commanded the morning and shown the dawn its place or took hold of the ends of the earth, till the wicked are shaken from its surface? Have you entered into the sources of the sea, or walked about in the depths of the abyss? Have the gates of death been shown to you, or have you seen the gates of darkness? Have you comprehended the breadth of the earth? Tell me, if you know all. Which is the way to the dwelling place of light, and where is the abode of darkness? I know, because I existed before them, and the number of my years is great! Have you entered the storehouse of the snow, and seen the treasury of the hail which I reserve for times of stress for the days of war in the battle? Have you fitted a curb to the plight of these or loosed the bonds of Orion? Do you know the ordinances of the heavens; can you put into effect their plan on the earth?" (Adapted by the author from the New American Bible.)

For example in psalm 8:3-4, the psalmist asks "When I consider your heavens, the work of your fingers, the moon and the stars, which you have ordained; what is man that you take thought of him and the Son of Man that you care for him?"

In Psalm 19 :1-7, the psalmist says "*The heavens declare the glory of God, in the firmament proclaims his handiwork. Day pours out the word today and night to night imparts knowledge; not a word nor a discourse whose voice is not heard; through all the earth their voice resounds and to the ends of the world, their message.*"

He has pitched a tent there for the sun which comes forth like the groom from his bridal chamber and, like a giant, joyfully runs its course. At one end of the heavens it comes forth, and its course is to their other end; nothing escapes from its heat."

Chapter 2
Did God have to create?

Marriage is a union of husband and wife. Their love for each other creates a child, an image of themselves.

So too with God the Father, and his love for his Son. The two divine beings produced the Holy Spirit. The comment I made about the Trinity is theologically incorrect. The Trinity has always existed, without beginning or end. The Nicene Creed, toward the end, Christians profess " belief in the Holy Spirit, the Lord, the giver of life, who *proceeds* from the Father and the Son. " So it is a divine happening that is ongoing, with no beginning or end.

Yet in their love for each other, the three persons of the Trinity, whose love for each other is certainly self satisfying, almost necessitates a further sharing of this marvelous, incomprehensible love that They share. This love each has for each other would lead them to create beings that could share in that love.

I find this analogy between the sacrament of marriage and the Trinity to elevate the vocation to the married life, on the lower level, as a divine gift. As love intensifies, so does the desire to share that love grow. This is not to take away from the mystics were celibate and experienced ecstasies that took them into the very heart of God. Saints like Teresa of Avila or St. Francis of Assisi or St. Padre Pio.

Reflection:

In the letter of St. Paul to the Ephesians, St. Paul writes the controversial chapter 5 about wife's being subordinate to their husbands as to the Lord., Most Catholic homilists tend to avoid this today, but there is an interesting part of this chapter that relates the relationship of the wife to the husband as a reference to Christ and his church.

Paul not only compares marriage to the relationship between Christ and the church but also expands on the latter. The passage abounds with instructions for marriage.

In this age, in the Catholic Church, a male dominated clergy has increasingly come under scrutiny by theologians.

Lay ministry is being done by very many women where the ordained clergy cannot possibly be engaged. Whether it be ministry to the sick, the homebound, catechetical leaders, pastoral assistants, or parish bookkeepers, women fill a much-needed void.

Protestant churches have long since dismissed the idea of a male-dominated clergy.

Chapter 3

Is this the only universe?

As we were looking back at Aristotle's concept of a prime mover, I began to realize that the concept of an eternal existing being was irrefutable.

If God is infinite, his creative power is also without limit. In understanding the concept of this universe, scientists today speak of the Big Bang theory. While many are given credit for developing this idea, I think the original concept came from a Jesuit priest.

Monsignor Georges Lemaître, a Belgian Catholic priest, was the originator of what would become known as the "Big Bang Theory"

Lemaître was born in 1894 in Charleroi, Belgium. As a young man he was attracted to both science and theology, but World War I interrupted his studies (he served as an artillery officer and witnessed the first poison gas attack in history). After the war, Lemaître studied theoretical physics, and in 1923 was ordained as an abbé. The following year, he pursued his scientific studies with the distinguished English astronomer Arthur Eddington, who regarded him as "a very brilliant student, wonderfully quick and clear-sighted, and of great mathematical ability." Lemaître then went on to America, where he visited most of the major centers of astronomical research. Later, he received his Ph.D. in physics from the Massachusetts Institute of Technology.

In 1925, at age 31, Lemaître accepted a professorship at the Catholic University of Louvain, near Brussels, a position he retained through World War II . He was a devoted teacher who enjoyed the company of students, but he preferred to work alone. Lemaître's religious interests remained as important to him as science throughout his life, and he served as President of the Pontifical Academy of Sciences from 1960 until his death in 1966.

In 1927, Lemaître published, in Belgium, a virtually unnoticed paper that provided a compelling solution to the equations of General Relativity for the case of an expanding universe. His solution had, in fact, already been derived without his knowledge by the Russian Alexander

Friedmann in 1922. But Friedmann was principally interested in the mathematics of a range of idealized solutions (including expanding and contracting universes) and did not pursue the possibility that one of them might actually exist. In contrast, Lemaître attacked the problem of cosmology from a thoroughly physical point of view, and realized that his solution predicted the expansion of the real universe of galaxies that observations were only then beginning to suggest.

By 1930, other cosmologists, including Eddington, Willem de Sitter, and Einstein, had concluded that the static (non-evolving) models of the universe they had worked on for many years were unsatisfactory. Furthermore, Edwin Hubble, using the world's largest telescope at Mt. Wilson in California, had shown that the distant galaxies all appeared to be receding from us at speeds proportional to their distances. It was at this point that Lemaître drew Eddington's attention to his earlier work, in which he had derived and explained the relation between the distance and the recession velocity of galaxies. Eddington at once called the attention of other cosmologists to Lemaître's 1927 paper and arranged for the publication of an English translation. Together with Hubble's observations, Lemaître's paper convinced the majority of astronomers that the universe was indeed expanding, and this revolutionized the study of cosmology.

A year later, Lemaître explored the logical consequences of an expanding universe and boldly proposed that it must have originated at a finite point in time. If the universe is expanding, he reasoned, it was smaller in the past, and extrapolation back in time should lead to an epoch when all the matter in the universe was packed together in an extremely dense state. Appealing to the new quantum theory of matter, Lemaître argued that the physical universe was initially a single particle—the "primeval atom" as he called it—which disintegrated in an explosion, giving rise to space and time and the expansion of the universe that continues to this day. This idea marked the birth of what we now know as Big Bang cosmology.

It is tempting to think that Lemaître's deeply-held religious beliefs might have led him to the notion of a beginning of time. The Catholic position was that God created *ex nihilo*, something out of nothing. Yet Lemaître clearly insisted that there was neither a connection nor a conflict between his religion and his science. Rather he kept them entirely separate, treating them as different, parallel interpretations of the world, both of which he believed with personal conviction. Indeed, when Pope Pius XII referred to the new theory of the origin of the universe as a scientific validation of the Catholic faith, Lemaître was rather alarmed. Delicately, for that was his way, he tried to separate the two:

"As far as I can see, such a theory remains entirely outside any metaphysical or religious question. It leaves the materialist free to deny any transcendental Being... For the believer, it removes any attempt at familiarity with God... It is consonant with Isaiah speaking of the hidden God, hidden even in the beginning of the universe."

In the latter part of his life, Lemaître turned his attention to other areas of astronomical research, including pioneering work in electronic computation for astrophysical problems. His idea that the universe had an explosive birth was developed much further by other cosmologists, including George Gamow, to become the modern Big Bang theory. While contemporary views of the early universe differ in many respects from Lemaître's "primordial atom," his work had nevertheless opened the way. Shortly before his death, Lemaître learned that Arno Penzias and Robert Wilson had discovered the cosmic microwave background radiation, the first and still most important observational evidence in support of the Big Bang.

When I was in college, I remember getting into a discussion with my friends about the possibility of universes, even multiple universes. Since I was a philosophy major, I remember the discussion about how many angels could dance on the head of a pin. Actually this was not such a silly discussion.

Although this discussion in scholasticism has some roots in Thomas Aquinas' "Summa Theologica", and that of Duns Scotus, modern philosophers have dismissed this argument as meaningless.

As a Catholic Christian, and an amateur astrophotographer, I don't think there's any end to physical space. One might think of this universe that we know, as a kind of bubble. Yet if it is just a bubble, what is beyond the bubble? My answer would be "more space." One can imagine many such bubbles, or, in other words, many universes. Could there be an infinite number of universes? Why not?

From my faith, I believe there is a Trinity made up of Father, Son, and Holy Spirit.

Could it not be that this triune God is active in this and other universes as well? The Vatican's chief astronomer has this to say:

The new president of the Vatican Observatory Foundation has said that it is only a matter of time before alien life forms are discovered, which will pave the way to questions about God's relationship to intelligent beings outside our planet.

Jesuit Brother Guy Consolmagno speculated that the general public will not be too surprised when life on other planets is eventually discovered, and will react in much the same way it did when news broke in the '90s that there are other planets orbiting far off stars.

Consolmagno, a planetary scientist who has studied meteorites and asteroids as an astronomer with the Vatican Observatory since 1993 has said that discovery of alien life will not prove or disprove the existence of God, but will pave the way to questions on salvation and how it relates to intelligent species.

Even Paul Francis in a sermon in May 2014, had this to say:

Pope Francis spent some time during a Mass at the Vatican on talking about alien life forms, and suggested that Martians, should they ever visit Earth, would be welcome to be baptized as well.

"If – for example – tomorrow an expedition of Martians came, and some of them came to us, here ... Martians, right? Green, with that long nose and big ears, just like children paint them ... And one says, 'But I want to be baptized!' What would happen?" the Roman Catholic Church leader theorized, as reported by Vatican Radio.

In his speech the pope focused on the question "Who are we to close the doors to the Holy Spirit?" He said that baptism is open to everyone, and reminded the audience of the words of Peter:

"If then God gave them the same gift He gave to us when we came to believe in the Lord Jesus Christ, who was I to be able to hinder God?"

Francis continued: "When the Lord shows us the way, who are we to say, 'No, Lord, it is not prudent! No, let's do it this way'... and Peter in that first diocese – the first diocese was Antioch – makes this decision: 'Who am I to admit impediments?' A nice word for bishops, for priests and for Christians. Who are we to close doors? In the early Church, even today, there is the ministry of the ostiary [usher]. And what did the ostiary do? He opened the door, received the people, allowed them to pass. But it was never the ministry of the closed door, never."

I believe that even in our Milky Way galaxy there are other intelligent races. Some may have had a progenitor or progenitors that were subject to a test, just as the Angels or Adam and Eve were. Some of these races may have passed the test and are living in a state of natural happiness. However, they are not in heaven yet. There is always going to be a choice before God closes the door on creation.

I believe that in the many universes that may exist and in the many life forms that most certainly do exist. Christ is doing the will of His Father and continues to give his life in a salvific act.

Chapter 4
Scripture and multiple universes

There is a quote in Scripture that might seem to contradict the concept of multiple universes or even alien life that is intelligent. It comes from the epistle to the Hebrews chapter 9, verses 24 through 28.

Christ did not enter into a sanctuary made by hands, a copy of the true one, but heaven itself, that he might not appear before God on our behalf. Not that he might offer himself repeatedly, as the high priest enters each year into the sanctuary with blood that is not his own; if that were so, it would have had to suffer repeatedly from the foundation of the world. But now once for all he has appeared at the end of the ages take away sin by his sacrifice. Just as it is appointed that human beings die once, and after this the judgment, so also Christ, offered once to take away the sins of many, will appear a second

time, not to take away sin but to bring
salvation to those who eagerly await him.

Does this quote contradict what we've been saying? I don't think it does. And the reason for it, is that it refers to the salvific act of Christ dying on the cross on our own planet in this , our galaxy, the Milky Way.

Yet there is another quote from the book of Wisdom, chapter 13 verses 1-9 that suggests that people in the past may have out of ignorance worshiped primitive gods. Let me quote:

> *For all men were by nature foolish who*
> *were in ignorance of God, and who from the*
> *good things seen did not succeed in knowing*
> *him who is and from studying the works did*
> *not discern the artisan;, but either fire, or*
> *wind, or the swift air, or the circuit of the*
> *stars, or the mighty water, or the luminaries*
> *of heaven, the governors of the world, they*
> *considered gods. Now if out of joy in their*
> *beauty they thought them gods, let them know*
> *how far more excellent is the Lord than these;*

for the original source of beauty fashioned them. Or if they were struck by their might and energy, let them from these things realize how much more powerful is he who made them. For from the greatness and the beauty of created things their original author, by analogy, is seen. But yet, for these the blame is less; for they indeed have gone astray perhaps, though they seek God and wish to find him. For they searched busily among his works, but are distracted by what they see, because the things seen are fair. But again, not even these are pardonable. For if they so far succeeded in knowledge that they could speculate about the world, how did they not more quickly find its Lord?

There are several other quotes in Scripture which speak of the end times or *parousia*. They occur in Matthew chapter 24, verses 15 to 22, in Mark chapter 13 verses 14 through 20, in Luke chapter 21 verses 20-24. In the second letter of Peter chapter 1 verse seven it reads:

The present heavens and earth have been reserved by the same word for fire, kept for the day of judgment and of destruction of the godless.

The book of Revelation, an apocalyptic work, contains many versions of the end times. Particularly chapters 21 and 22 speak of the new heaven and the new earth. It has this to say:

> " *Then I saw a new heaven and a new earth. The former heaven and the former earth had passed away, and the sea was no more. I also saw the holy city, the new Jerusalem, coming down out of heaven from God, prepared as a bride adorned for her husband. I heard a loud voice from the throne saying, 'Behold God's dwelling is with the human race. He will dwell with them and they will be his people and God himself will always be with them as their God. He will wipe away every tear from their eyes, and there shall be no more death, or morning, wailing, or pain, for the old order has passed away.'"*

In chapter 22 it reads:

"Then the angel showed me the river of life-giving water, sparkling like crystal, flowing from the throne of God and of the Lamb down the middle of its street. On either side of the river grew the tree of life that produces fruit 12 times a year, once each month; the leaves of the trees serve as medicine for the nations. Nothing accursed will be found there anymore. The throne of God and of the Lamb will be in it, and his servants will worship him. They will look upon his face, and his name will be on their foreheads. Night will be no more, nor will they need light from lamp or sun, for the Lord God shall give them light, and they shall reign forever and ever."

So what are we to make of all of these Scriptures? Will the universe as we know it today actually be done away with? This is a question which is the subject of scholarly discussion. The Catholic Church in its pronouncements on science has been wrong before. The case of Galileo and his observations of the four moons of Jupiter and the phases of Venus that he saw through his telescope, earned him the scorn of the papacy and house arrest. Apocalyptic writings present a sense of urgency, almost a sense of fear and dread.

In the nearest galaxy to the Milky Way, lies the Andromeda galaxy.

The Andromeda galaxy, also known as Messier 31, M31, or NGC 224, is a spiral galaxy approximately (2.5 million light years). It is the nearest major galaxy to the Milky Way.Being approximately 220,000 light years across, it is the largest galaxy of the Local Group, which also contains the Milky Way, the Triangle galaxy, and about 44 other smaller galaxies.

The Andromeda Galaxy is the most massive galaxy in the Local Group as well, containing at least twice the number of stars in the Milky Way, which is estimated to have 200–400 billion stars.

Now in the Andromeda galaxy, would it not be reasonable to assume that there are other planets similar to ours? Astronomers have discovered two new alien worlds a bit larger than Earth circling a nearby star.

The newfound exoplanets known as HD 7924c and HD 7924d, are "super Earths" with masses about 7.9 and 6.4 times greater, respectively, than that of our home planet, researchers said. The planets orbit the star HD 7924, which lies just 54 light-years from the sun — a mere stone's throw considering the size of the Milky Way, which is on the order of 100,000 light-years wide.

The discovery brings the number of known planets in the HD 7924 system top three. (Another super Earth, called HD 7924b, was spotted there in 2009.) HD 7924b, HD 7924c and HD 7924d all lie closer to their host star than Mercury does to the sun. They complete one orbit in five, 15 and 24 days, respectively, researchers said.

"The three planets are unlike anything in our solar system, with masses seven to eight times the mass of Earth and orbits that take them very close to their host star," study co-author Lauren Weiss, a graduate student at the University of California, Berkeley,

The research team discovered HD 7924c and HD 7924d using three different ground-based facilities — the Automated Planet Finder (APF) Telescope at Lick Observatory in California, the Keck Observatory in Hawaii and the Automatic Photometric Telescope (APT) at Fairborn Observatory in Arizona. (Keck also found HD 7924b in 2009.)

The research team, which was led by University of Hawaii (UH) graduate student BJ Fulton, used the combined observations of the three telescopes to detect tiny wobbles in the star HD 7924 caused by the gravitational pull of the two newfound planets, and then to verify the worlds' existence.

Astronomers have confirmed more than 800 planets beyond our own solar system, and the discoveries keep rolling in.

The APF Telescope was recently revamped to make it fully robotic, and it now searches the skies for exoplanets without human oversight — a key milestone in the ongoing exoplanet hunt, researchers said.

"This level of automation is a game-changer in astronomy," said co-author Andrew Howard, an astronomer at UH. "It's a bit like owning a driverless car that goes planet shopping."

Astronomers first found planets orbiting another star in 1992, and the exoplanet tally has now risen to nearly 2,000. More than half of these alien worlds have been discovered by NASA's Kepler space telescope ,which launched in March 2009.

You have to wonder whether or not these exoplanets may have some form of life, even intelligent life.

These discoveries were made as late as May of 2015. We are just beginning to explore the billions of exoplanets planets there are in all of the galaxies in the known universe.

In the universe as we know it today, astronomers can use the Hubble space telescope to review galaxies near the end of the observable universe by examining a very tiny portion of the sky, hunting up the number of galaxies visible in that region, and then multiplying that number to account for all the regions of the sky. This is how astronomers estimate how many galaxies they are in the observable universe. A 2013 study indicated there are 225 billion galaxies in the observable universe.

The galaxies that the Hubble telescope sees have morphed and expanded because light travels at a fixed rate of speed…. 186,000 mi./s. So, some of them that are seen by Hubble may not exist in the visible spectrum. We spoke earlier about the possibility that this universe as we know it, might only be one of a countless number of universes that we know nothing about.

When the writers of Scripture say that there will be a new heaven and a new earth, as a Catholic I can believe that. But does that mean that on this earth when Christ comes at the second coming, will that mean that all of God's creation will be done away with?

If intelligent life were discovered on another planet, what would this do to Christianity (and other world religions)?

The history of Christianity has rested on the assumption that rational beings with spiritual souls residing in earthly bodies are found on this world only. The recent discoveries of exoplanets outside our solar system leads the scientist and theologian to conclude that it is almost unthinkable that intelligent life does not exist anywhere but Earth. There are billions of galaxies with billions of suns with countless billions of planets in the physical universe. The odds that Earth alone has intelligent beings is slim indeed.

Judeo-Christian scripture mentions intelligent, body-less beings. According to the Bible, there are 94 references to "angels," spiritual beings who often take a temporary "bodily" form to serve as messengers to mankind from God. Medieval theologians attempted, from the Old and New Testaments, to reason that there was a schema of three hierarchies of angels, with each hierarchy containing three Orders or Choirs. In rank, from highest to lowest were the Seraphim, Cherubim, Thrones, Dominions, Virtues, Powers, Principalities, Archangels, and Angels. However, modern scripture scholars find their interpretation of Ephesians 1:21 and Colossians 1:16 to be very tentative and ambiguous sources. In the Old Testament, there are 59 references to Cherubim, with their first appearance coming in Genesis 3:24 after God expelled Adam and Eve from the garden of Eden, stationing the cherubim and the fiery revolving sword, to keep them from returning . Angels are thought to have been "tested" as were Adam and Eve; however, their test was thought to be a once-onlychoice between God or Lucifer.

But, let us return to beings not gifted with the intellectual gifts angels are presumed to have. What if earth-like beings, similar to humans, exist? If they descended from an "Adam and Eve" on their planet, did their first parents sin and pass that original sin to their offspring? If so, was there a need for a redeemer, another Christ? If the experience of the human race on Earth is any guide, most probably. Could the Second Person of the Trinity be the Redeemer on other planets? Why not?

What if their "Adam and Eve" did not sin? Would their descendants be tested somehow? I don't think God makes robots. Free will and choice are part of his plan.

There also are other considerations…what if intelligent beings had little or no physical dimensions … ETs, for example, or had "super-brains" with vast intellectual capabilities. Again, each race would be subjected to God's appropriate test.

What if intelligent life lived within a microscopic universe, with molecules the size of solar systems in their view …. What if, what if … why limit God, who has infinite power to create in any physical or spiritual dimension He chooses?

Cosmologists think they have evidence that could help prove that our universe is just one of many others. So far, the "multiverse theory" has been a controversial notion without hard evidence to support it. That may now change.

Cosmologists studying the highly detailed data from the Planck Telescope, a space telescope designed to measure the universe's cosmic microwave background radiation, say they have evidence that our universe is one of many others in existence.

The cosmic microwave background radiation (CMBR) is known as the "afterglow" of the big bang. It fills the background of the universe almost uniformly, but highly sensitive instruments, such as the Planck Telescope, can detect minute variations in this apparent uniformity.

What the Planck spacecraft has revealed is that the CMBR is far from uniform. In fact, the radiation shows up stronger in the southern half of the map versus the northern half. There is also a large "cold spot" in the southern sky, which modern physics can't explain. Until now, perhaps.

Laura Mersini-Houghton, theoretical physicist at the University of North Carolina at Chapel Hill, and Richard Holman, professor at Carnegie Mellon University, both theorized in 2005 these anomalies could be caused by the pull of other universes outside our own.

However, these notions were mostly

speculation until recently when new data from the Planck Telescope provided a much more detailed map of the radiation. Now the pair believes the data provides the first solid proof of multiverse theory.

To date, their hypothesis is the only one that can explain the stronger radiation in the south and the cold spot in that hemisphere.

The Multiverse theory, if true, could explain a number of things including what caused the Big Bang.

Indeed, prominent atheists such as Stephen Hawking have proposed the multiverse theory as an alternative explanation for the story of Biblical creation and why things are the way they are. However, their use of the theory does not mean that creation occurred in any particular way or that there can't be other universes besides ours.

It is entirely plausible to accept the Big Bang theory and multiverse theory while maintaining a steadfast belief in God. God could have used any means He wished to create.

The analysis of the Planck Telescope map is certainly interesting, but a great deal further research will be required before we can discern if other universes exist and have somehow exerted a detectible influence on ours. The conclusions are already controversial in the scientific community and scientists will debate them for a long time to come.

So, in conclusion, it is my opinion that Christianity would stand up to any discoveries of rational beings on other planets. Our Christ could be known by other names in other planets' histories. All creation comes from the Father, to be sure. Redemption would depend on the success or failure of the races of that particular planet to embrace God's will to "choose good and avoid evil."

Chapter 5
Pierre Teilhard de Chardin

    Pierre Teilhard de Chardin was a French philosopher and Jesuit priest who trained as a paleontologist and geologist. During his lifetime many of Teilhard's writings were condemned. Teilhard's views on original sin and consequently many of his works were censured by the Catholic Church throughout his lifetime. Although a number of theologians deny that his writings and thought were heterodox, a closer examination is likely to yield a different conclusion. Regardless of this, many have recently considered him to be a visionary not only for his creative work on the relationship between science and theology but also for the advent and progression of the internet, globalization, eco-theology and contemporary trans-humanism. Given the wide interest in these subjects today, it is hardly surprising that contemporary writers are recycling Teilhard's ideas. Even some prominent theologians have embraced rather uncritically much of Teilhard's thought. Yet despite his intriguing ideas, his works remain fraught with scientific, theological and philosophical difficulties. If Teilhard's thought is going to be presented to a new generation, it should be done so in an honest and objective manner

In Chapter 2 of the *Spirit of the Liturgy*, *(2009)* Benedict XVI describes how Teilhard's theological vision of Christ is central to the Christian liturgical and Eucharistic experience:

> *"And so we can now say that the goal of worship and the goal of creation as a whole are one and the same—divinization, a world of freedom and love. But this means that the historical makes its appearance in the cosmic. The cosmos is not a kind of closed building, a stationary container in which history may by chance take place. It is itself movement, from its one beginning to its one end. In a sense, creation is history. Against the background of the modern evolutionary world view, Teilhard de Chardin depicted the cosmos as a process of ascent, a series of unions. From very simple beginnings the path leads to ever greater and more complex unities, in which multiplicity is not abolished but merged into a growing synthesis, leading to the*

"Noosphere", in which spirit and its understanding embrace the whole and are blended into a kind of living organism. Invoking the epistles to the Ephesians and Colossians, Teilhard looks on Christ as the energy that strives toward the Noosphere and finally incorporates everything in its "fullness". From here Teilhard went on to give a new meaning to Christian worship: the transubstantiated Host is the anticipation of the transformation and divinization of matter in the christological "fullness". In his view, the Eucharist provides the movement of the cosmos with its direction; it anticipates its goal and at the same time urges it on."

As I have noted, Teilhard's views on original sin and consequently many of his works were censured by the Catholic Church throughout his lifetime. Although a number of theologians deny that his writings and thought were heterodox, a closer examination is likely to yield a different conclusion.

Aside from not providing a satisfactory solution to the problem of evil since the problem remains regardless of the process by which God chose or chooses to create, Teilhard does not substantiate his claim that God can only create solely through evolution. Why limit a sovereign God in such a way? Even if God were to *solely* create through evolution, there is no reason to think that he could still not be intimately involved in the process, perhaps acting at the quantum level leaving such an involvement ambiguous.

It is clear that Teilhard has an agenda to reconstruct the traditional conception of God, one in which eventually even God must bow down to the process of evolution and goes from being the evolver to part of the evolved. Teilhard attempts to do this through the use of science. Of course science cannot fully adjudicate such overarching metaphysical and theological issues. At best science can lead us in a particular direction when incorporated in an overall philosophical argument, but not solely on its own since by definition it would cease to be a "scientific" explanation.

Teilhard has also a peculiar vision of Christ in lieu of his views on evolution. He sees Christ as an evolving Christ, much as his vision of God becoming part of the evolutionary process.

De Chardin's two main works are *The Phenomenon of Man* and *The Divine Milieu.*

When I was in the seminary at St. Mary's University in Baltimore Maryland in the early 60s, Chardin's works were forbidden reading. A decree of the holy office dated June 30, 1962 under the authority of Pope John XXIII warned that " it is obvious that in philosophical and theological matters the works of Teilhard are replete with ambiguities and rather with serious errors which offend Catholic doctrine. That is why this holy office urges all bishops superiors and directors of seminaries to effectively protect, especially the minds of the young, against the dangers of his works."

Further, in 1963 the Vicariate of Rome ruled under the name of Pope Paul VI who had just become Pope in 1963 in a decree dated 30th of September required that Catholic booksellers in Rome should withdraw from circulation the works of Chardin together with those books which favor "his erroneous doctrines."

Needless to say, I didn't read the books. However, now that I have studied theology in depth, I see in some of his teachings a common chord with what I believe.

Teilhard's view may be summed up in this way: *All of creation, whether this universe, or the many many other universes that may exist, will all come together in the Omega point which is God and his person Jesus Christ.* It's here where I part company with this scholar.

I believe that the scriptural references that we have seen refer to this area of the universe.

We are myopic, I think, to imagine that we are the only intelligent beings in a physical universe. I am also convinced that there are multiple, even countless universes where the Trinity is active. These universes, which of course we have no proof of, certainly, in my opinion, do exist.

I believe that Christ will be obeying his Father in other civilizations of time and space for as long as the Father wills it.

Chapter 6

UFO's and extraterrestrials

Countless books, movies, television shows and radio programs have been written and produced describing mankind's encounters with alien beings from other planets. In fact it is so commonplace now that our very language and culture has been affected by these concepts. Nonetheless, I am sure you have heard very little if anything mentioned by the Catholic Church or any other denomination for that matter. In this brief article I wish to touch upon the concept of extraterrestrials, their existence or nonexistence and if they did existence what this would mean for Roman Catholics.

It would appear from all the evidence that something is out there. Countless men and women throughout the world having had no contact with each other have nonetheless reported encounters with alien spacecraft, aliens themselves or being abducted by aliens into their spacecraft. Could all those people be lying? Are they all delusional? Evidence seems to indicate the contrary. I would say many of us, if we have not had our own experience of these "Outworlders", know of someone within our immediate sphere of influence who has had an alien encounter of one kind or the other.

There are several kinds of "encounters." One kind are sightings of some kind of unexplained lights in the sky. . A whole industry has risen up to accommodate the plethora of sightings every year by simple uneducated rural folk to sophisticated doctors, lawyers, military officers, police. and even astronauts.

These sightings are many and various, occurring in virtually all inhabited continents. Another kind involve "evidence" left by UFOs, such as burned ground, etc. The more extreme "encounter" are abductions. These are the designated types of encounters with extraterrestrials gleaned from the thousands upon thousands of reported encounters.

So what does all this mean? Do Unidentified Flying Objects exist or are they simply the product of some mass hallucination? Are there extraterrestrial beings with enough intelligence to have created interstellar travel or are we alone in the universe?

I think that the witnesses of these events are not crazies or candidates for psychiatric treatment.

It is the general assumption (not mine) of those who have studied this apparently universal phenomena that these objects and their occupants are extraterrestrial; i.e., from another planet in a solar system within our galaxy, the Milky Way. The theory is that these creatures have reached a higher level of evolution and their intelligence is as superior to ours as our intelligence is to gorillas. They have developed the possibility of space travel in vehicles that are capable of traveling through space at speeds that we have only fantasized about in science fiction stories.

A theory less popular then the one mentioned above but still posed by some people is that these creatures actually originate on Earth and come from under the sea. They are the survivors of the first age of the world, which was destroyed through natural disaster, and they are now reappearing to warn mankind of impending dramatic Earth changes.

Another theory says they are actually from our own solar system. The story goes that there is a twin planet "Earth" which is perfectly opposite the Earth on the other side of the sun and therefore undetectable. I think this is whacko.

Interdimensional travel is yet another theory. This theory claims that these creatures come from another dimension, a parallel universe as it were, which actually occupies the same space as we do but is in another dimension unseen and with which we are unable to communicate or even know of its actual existence.

There is also the theory that these are time travelers from our own future. According to this theory our future ancestors are the products of eons of evolution. They have returned to our time to investigate their origins and to help eventually advance the human race toward greater evolution.

In actual fact, every one of these theories falls flat. Incredibly, with all the sightings and experiences of possibly millions of people since the late 1940's there is absolutely no hard evidence (of which we are aware) that indicates that these beings and the lights they produce in the sky are from another planet, another dimension, another time or from a lost civilization of the past.

Let me explain. Actually we are surrounded by millions of powerful beings which are purely spiritual and which are both good and evil. We call these beings, angels or demons depending upon which side of the eternal fence we are speaking. They surround us. They are in the very air, the sky the earth. They exist on another plane of existence than we do yet they are in constant contact with this world. They have powers beyond our comprehension and if one presented himself as God to us, unless we had the discernment of spirits we could very easily believe he was God or at least a god. Thus in ancient times demons appeared to men as gods demanding their worship and sacrifices. This was done in an effort to lead men into fear and away from the light of truth. Enslaving millions of human beings they would perform wonders to convince them of their divinity and nearly the whole world was in darkness.

As science and technology moved Western civilization toward new levels of affluence and secularism the worship of idols and a pantheon of gods became less and less prevalent. With the triumph of Christianity idolatry was all but eliminated except in a few cases. (Due to African influence idolatry and ritual witchcraft still exist today in false religions such as Voodoo and Santeria.) In the East, however, many millions of superstitious Hindus are still immersed in idolatry.

So what am I trying to say?. This UFO phenomenon witnessed by millions of people all over the world might simply be the activity of demons, or of forces we simply do not understand.

After all how many times do aliens have to probe the sexual organs of people? Either they are stupid, but yet they're supposed to be super intelligent, so I think it's quite possible that they are demonic.

Now it must be noted that the reason for demonic attacks upon human beings are three fold: First, to distract our minds from God and the salvation of our own souls by turning us to focus on them and their activity: Second, to create a sense of terror, panic, emotional distress, and finally despair in the love and goodness of God: Third, to damn the soul of the one possessed and to influence toward damnation those around them. In cases of possession there can be a fourth reason and that is to attack the priest who will be called in to exorcise the demon.

In what way do the activities of demons bear similarity to those experienced by people who witness UFO phenomenon? Well the first and most obvious similarity is that focusing on UFOs and their occupants distracts us from striving for union with God and the salvation of our souls. If you have ever been around those who are interested in UFO phenomenon you will find that they become obsessed with the idea of extraterrestrials. They are so focused upon this sensational idea that they can hardly think of anything else.

Another experience that indicates demonic activity is that these aliens inspire terror and fear within those who have had contact with them. They describe a feeling of terror similar to that of a caged or cornered animal; so primal, in fact, that it is very difficult for them to describe the panic and hopelessness they feel. The demons feed off of our negative emotions and so it is their modus operandi to inspire as many negative emotions as they can within their victims.

If you read accounts of abduction you will find that many times the memories of these abductions are buried so deeply within the subconscious of the victims that it takes hypnosis to bring it back to their conscious minds. The experience when relived causes the victim to feel the same terror they originally felt. Some would say that the very use of hypnosis to stimulate the subconscious borders on the occult and plays right into the hands of the demon.

The abductees often suffer from ailments that our medical doctors have not seen nor do they have any idea of how to treat or cure them. Often they have scratches on their bodies that will not heal (another similarity to those who have encounters with the demonic).

When a so-called alien enters into a room to abduct someone, things in the room begin to move on their own accord. The person is paralyzed and cannot move or even call out in fear. When taken to their "alien spacecraft" they find themselves powerless as these creatures begin to experiment upon them. All calculated to inspire fear and terror.

Could extraterrestrials exist?

The answer to the above question is an emphatic yes. There is no reason from a Catholic point of view that we have to reject without possibility the existence of creatures like ourselves in other parts of the universe. Nonetheless, we can only accept this possibility within certain parameters. For instance, we must grant the fact that they have souls and are also made in the image and likeness of God just as we are. Therefore, they can be in only two states of existence. They are either fallen or unfallen. In other words, because they would be rational creatures made in the image and likeness of God they would also have to have been given the choice between embracing the One True God or the god of self. If they chose God they would have remained unfallen and would never have suffered the terrible twisting of their nature that original sin brings. They, more than likely, would have no desire or need to go beyond their own boundaries. Their technology, if any, would be simple and practical. They would be the sovereigns of their own world and content within themselves. These humans (in whatever shape or form) would have no need to explore the universe. Finally since they would be unfallen they would not need the redemptive grace of Christ to free them from bondage to the Devil. They would essentially be what we would have been had we not fallen. This of course raises any number of questions that would be further theological speculation and which I do not have the space here to pursue. Suffice it to say these beings would not be in contact with us unless for some reason God Himself sent them to us as a grace. If this were to happen they would not be secretive but would present themselves visibly without stealth or subterfuge.

If there are others "out there" and they have a fallen nature then we must be very careful. For they may act exactly as these reports indicate. They would be under the influence of the Devil and would not have had the grace of Christ to lead them to the virtue for which God originally made them. If man can be as inhuman as he is to his own even with the redemptive grace of Christ to mitigate the effects of original sin, how much more would fallen humans from another planet be possessed of a corrupt nature not having had the opportunity of Christ's redemptive grace to be given to them or their society? They could make Hitler and Stalin look like Saints.

It is important to remember that according to Catholic theology, the Second Person of the Blessed Trinity became man suffering death "once and for all..." to save all those who have human souls from the damnation our sins deserve.

In truth book upon book could be written based solely upon theological speculation concerning the existence of intelligent extraterrestrials and their place within the providence of God. As long as we realize that their souls would have to abide by the same spiritual principles by which we ourselves must abide there are thousands of different ways we could speculate about this matter.

Until proven otherwise I'm sticking to my previous analysis of this phenomenon and remain convinced that these are merely demonic manifestations designed to distract and disturb those who encounter them.

You've probably heard of SETI, which is an acronym Search for Extra Terrestrial Intelligence. The following is from the website www.seti.org/

This is an incredibly exciting time for the SETI Institute. The number of verified planets outside of our solar systems grows rapidly, and includes several that may have liquid water on their surfaces. At the same time, we are learning that life can survive in amazing places, even in lakes sealed beneath the Antarctic ice. These and other recent developments virtually assure the existence, and ultimate verification, of life beyond Earth. Recently NASA Chief Scientist Ellen Stofan agreed, predicting "I think we're going to have strong indications of life beyond Earth within a decade, and I think we're going to have definitive evidence within 20 to 30 years." The SETI Institute is at the forefront of many of these discoveries. As the only organization that addresses the full range of disciplines related to understanding and explaining the origin and nature of life in the universe, we search for answers to critical questions such as:

- How did life begin on Earth?
- Where/when/how did it overcome bottlenecks?
- Does it exist elsewhere?
- Are there other technological life forms?
- Can we survive our own technological adolescence?
- Is there a long future for life on Earth?

The answers to these and related questions are critical for informing some of the most important decisions mankind will make in the next 50 years.

Unfortunately, it has become increasingly difficult for basic research institutions like our Institute to rely on government funding in the United States. Increasingly, we are dependent on visionaries such as you to support our programs. When you invest in the SETI Institute, you join leaders such as Dave Packard, Paul Allen, Bill Hewlett, Gordon Moore, and Franklin Antonio, all of whom have been strong supporters.

After nine years en route, the New Horizons spacecraft is nearing Pluto for the July 14 flyby. We have many ideas to enhance our public and student engagement based on the expertise of our scientists who are participating in this mission, but we lack the funding to execute on them.

The Institute's NASA Astrobiology Team uses innovative, autonomous rovers in the high lakes of the Andes to simulate landers that will float in Titan's ethane lakes. This team strives to better understand planetary responses to rapid climate change. We have multiple opportunities for independent studies by postdoctoral fellows using these data, but we must find funds to support them.

The Allen Telescope Array is being upgraded with more sensitive radio receivers, capable at working at even higher frequencies, to improve the search for other technological civilizations. This improved sensitivity is like building more telescopes, making the search even more effective.

Chapter 7

What is meant by the Book of Revelation's "a new heaven and a new earth"?

Just as the life on earth is dependent upon the energy from the sun for sustenance, so does the fate of our solar system hinge on the sun' survival. Our sun which is classified as a yellow drwarf is a middle-aged star that's approximately 5,000,000,000 years old. As a main sequence star with a finite lifespan it will eventually die. This and will occur following the depletion of the last of the hydrogen forged in its core.

But what exactly will happen when it uses the last of its hydrogen? Much of the following information was taken from a writing by futurism.com/Wg4mP

Well, when this occurs, the core of the sun will shrink under its own gravity and become so dense that the helium atoms will begin to collide to form carbon and oxygen atoms. The collisions of said elements will churn out more energy than what the sun currently does. This is because the current amount of energy is dependent on the sun's fusion of hydrogen into helium, and hydrogen and helium are the lightest elements. The extra energy will prove to be the beginning of the end for the earth. The one the core will increase in temperature causing it to swell to hundreds of times its present size, changing its status from a yellow dwarf to a red giant, which will certainly be the end of the two innermost planets of our solar system, Mercury and Venus.

The fate of our home planet might be up for grabs. Many scientists speculate that our earth will become a blackened blob, and that it will be consumed by our sun. Some others claim that the Earth will be pushed out of orbit, away from the sun; however, regardless of whether Earth survives total incineration during the initial increase in the sun's surface area, and will no longer be habitable for humans. The oceans will boil and evaporate, the atmosphere will be blown away, lost to space forever, while all of the lush vegetation, along with any surviving ecosystems, will be destroyed. All that will be left is a barren wasteland not fit for humans.

The outer planets, those that are located beyond the orbit of Mars, may have another fate . Some of the moons of Jupiter and Saturn e are known to host frozen bodies of liquid water under its icy surfaces. Europa is believed to contain more water on it then there is water in all of the oceans lakes and streams on earth.

Thanks to the huge expansion of the sun, places in the outermost regions of our solar system will have a few million years left/ Think of it as a short springtime after a 10 billion year winter, one last golden era before the sun sets for the final time and perpetual darkness takes over.

Meanwhile, on the sun, the helium burning reaction will produce strong solar winds As the material withdraw from the sun's surface and into the surrounding region, it will carry off some of its remaining hydrogen in its outermost layers, forming a brilliant planetary nebula. The resulting nebula would be visible for thousands of years to any civilizations existing within a few hundred light-years of our former stellar neighborhood.

Eventually, that too will end when there isn't enough pressure at the core to keep the process of helium fusion going. What's left of our sun will inevitably contract under its own gravity and become a much more compact, dense star that will be left over from the earlier nuclear fusion. At this point, our once life-giving sun will be a white dwarf and the king of the dead solar system. A white dwarf will eventually cool over a few hundred million years, with the remaining mass being jammed into a sphere approximately the same size as earth, but much more dense. Eventually, this energy will dissipate, leaving behind a stellar corpse called a black dwarf, effectively ending the journey of life in our solar star system.

So are the end times and the new heaven and new earth referring only to our solar system?

Even in our own galaxy, there are stellar explosions that are going on. One of them Eta Carinae, the galaxy's biggest, brightest, and perhaps most studied star after the sun, has been keeping a secret. It's giant outbursts appear to have been driven by an entirely new type of stellar explosion that is fainter than the typical supernova and does not destroy the star there is a class of stellar explosions going off in other galaxies for which we still don't know the cause, but Eta Carinae is a good example. This outburst occurred in 1843 and was, in fact, an explosion that produced the fast blast wave similar to, but less energetic than a real supernova.

If we think of the 290 billion galaxies or so in the known universe, there are countless explosions going on all the time. Are these the end times for other civilizations?

Chapter 8
Opening up the mind

As I write this book, another of the iterations of the Star Trek movies is playing in movie theaters. The original Star Trek movie was seen by some theologians as a parody of the forces of good and evil in the world There is the force of darkness and the force of light and this is the story of Scripture itself.

Yet when we look at the distances that are portrayed in these movies, (science fiction to be sure), it is hard to imagine that our civilization on this planet will ever have the technology needed to travel light years in matter of minutes.

There are visionaries out there, however who suggest that we will be able to mine asteroids by 2025. Planetary resources deployed its first spacecraft from the international space station in July 2015, and the Washington-based asteroid mining company aims to launch a series of increasingly ambitious and capable probes over the next few years.

The goal is to begin transforming asteroid water into rocket fuel within a decade, and eventually to harvest valuable and useful platinum group metals from space rocks. Personally I think this is a far-fetched idea.

I am not a literalist when it comes to Scripture, but when God said fill the earth and subdue it, I don't think he was talking about mining asteroids!

I hope this book, short as it is, has given the reader a perspective not only on how immense God's creation is, but how lovely He is for his creatures.

This is the miracle that astronomy has opened for me.

www.ingramcontent.com/pod-product-compliance
Lightning Source LLC
Chambersburg PA
CBHW071625170526
45166CB00003B/1204